Unraveling QBism

—∭—

A Deep Dive into Quantum Bayesianism

With ONE MATH Equation

Dr. Sanjay Basu

Acknowledgment

I dedicate this book to my dearest wife, Soma Chakravorti, for her unwavering love, support, and belief in me throughout this journey. Her presence has been a beacon of light, illuminating the path through the complexities of life and quantum mechanics. As the observer in my world, she has given meaning to my own quantum superpositions and helped me navigate the entanglements of our shared existence. I thank her for being the constant source of inspiration, understanding, and encouragement. Her love has transformed the uncertainties of life into a beautiful dance of possibilities. Here's to continuing our journey together, exploring the mysteries of the universe, hand in hand.

Table of Contents

Forewords

Quantum mechanics has been a cornerstone of modern physics for over a century. Since its inception, it has challenged our intuition, rewritten the rules of reality, and provided a framework for understanding the microscopic world. Despite its immense predictive power, the interpretation of quantum mechanics remains a subject of debate and confusion.

This discourse aims to provide a comprehensive introduction to Quantum Bayesianism, or QBism, one of the most recent and intriguing interpretations of quantum mechanics. We will delve into the core principles of QBism, analyze how it addresses longstanding quantum paradoxes, and explore its implications for the future of quantum science.

Join us on a journey through the fascinating world of QBism, where we explore the implications of viewing quantum mechanics through a personal, subjective lens. With a thorough analysis of the foundational concepts, real-world applications, and potential future developments, this book aims to provide readers with a solid understanding of QBism's role in shaping our understanding of the quantum world.

1
Introduction to Quantum Mechanics

1.1. A Brief History of Quantum Theory

Quantum mechanics was born out of the need to explain phenomena at the atomic and subatomic scales that classical mechanics could not. The development of quantum theory began in the early 20th century with several key breakthroughs:

In 1900, Max Planck introduced the concept of quantization of energy to explain black-body radiation. He proposed that energy could only be emitted or absorbed in discrete packets called quanta.

In 1905, Albert Einstein built upon Planck's work, explaining the photoelectric effect by proposing that light itself was quantized into particles known as photons.

In 1913, Niels Bohr developed a model of the hydrogen atom that incorporated quantization, explaining the discrete energy levels of the electron orbiting the nucleus.

The 1920s saw the development of the two key mathematical formulations of quantum mechanics:

Werner Heisenberg, Max Born, and Pascual Jordan developed

matrix mechanics in 1925, which treated observables as matrices operating on quantum states.

In 1926, Erwin Schrödinger introduced wave mechanics, describing the behavior of particles as wave-like entities governed by a wave equation.

These formulations were later shown to be equivalent, forming the basis of modern quantum mechanics.

1.2. The Core Principles of Quantum Mechanics

Quantum mechanics is built upon several key principles that distinguish it from classical mechanics:

Wave-particle duality: Particles exhibit both wave-like and particle-like behavior, depending on the context of the observation.

Quantum superposition: Quantum systems can exist in a linear combination of multiple states simultaneously until a measurement is made.

The uncertainty principle: There is a fundamental limit to the precision with which certain pairs of physical properties (e.g., position and momentum) can be known simultaneously.

Quantum states and wave functions: The state of a quantum system is described by a complex mathematical function known as the wave function, which contains all the information about the system.

Probability and the Born rule: The outcome of a quantum measurement cannot be predicted with certainty, but only with probabilities given by the square of the absolute value of the wave function.

1.3. The Measurement Problem

The measurement problem is one of the central puzzles in quantum mechanics. It arises from the apparent contradiction between the continuous, deterministic evolution of the wave function according to the Schrödinger equation and the discontinuous, probabilistic outcomes of quantum measurements.

The issue is exemplified by the famous Schrödinger's cat thought experiment, which illustrates how a quantum superposition can lead to macroscopic superpositions that seem absurd from a classical perspective. The measurement problem has led to the development of various interpretations of quantum mechanics, each offering different solutions.

1.4. Entanglement and Nonlocality

Quantum entanglement is a phenomenon in which two or more particles become correlated in such a way that the state of one particle is dependent on the state of another, regardless of the distance between them. Entanglement is a unique and non-intuitive aspect of quantum mechanics that has no classical analog.

The implications of entanglement were brought to light in 1935 by Albert Einstein, Boris Podolsky, and Nathan Rosen in the famous EPR paradox. They argued that quantum mechanics was incomplete because it seemed to allow for "spooky action at a distance," where the measurement of one particle instantaneously affected the state of another, distant particle.

In 1964, John Bell derived a set of inequalities that could be used to experimentally test whether entanglement could be explained by local hidden variables. Subsequent experiments have consistently violated Bell's inequalities, providing strong evidence for the nonlocal nature of quantum.

2
Quantum Interpretations: The Quest for Meaning

2.1. The Copenhagen Interpretation

The Copenhagen interpretation, developed by Niels Bohr and Werner Heisenberg in the 1920s, was the first widely accepted interpretation of quantum mechanics. It emphasizes the role of observation and measurement, asserting that only the results of measurements have definite values. Key aspects of the Copenhagen interpretation include:

Wave function collapse: Upon measurement, the wave function of a quantum system collapses into a definite state, chosen probabilistically according to the Born rule.

Complementarity: Quantum systems exhibit both wave-like and particle-like properties, but these complementary aspects cannot be observed simultaneously.

The observer's role: The act of measurement, involving an observer or measuring device, plays a crucial role in the manifestation of physical properties.

Despite its historical significance, the Copenhagen interpretation has been criticized for its lack of clarity regarding the nature of wave function collapse and the role of the observer.

2.2. Many Worlds Interpretation

The Many Worlds Interpretation (MWI), first proposed by Hugh Everett III in 1957, offers a radically different perspective on quantum mechanics. It postulates that every possible outcome of a quantum measurement actually occurs in a separate, non-communicating branch of the universe. Key aspects of the MWI include:

Universal wave function: There exists a single wave function describing the entire universe, which evolves deterministically according to the Schrödinger equation.

Branching: When a measurement is made, the universe splits into multiple branches, each corresponding to a possible outcome.

No collapse: Unlike the Copenhagen interpretation, the MWI does not involve wave function collapse.

Critics of the MWI argue that it introduces unnecessary complexity and fails to provide a clear explanation for the apparent randomness of quantum outcomes in our observed universe.

2.3. Bohmian Mechanics

Bohmian mechanics, also known as the de Broglie-Bohm theory or pilot-wave theory, was proposed by David Bohm in 1952 as an alternative to the Copenhagen interpretation. It provides a deterministic and nonlocal hidden-variable theory that restores a particle-based ontology to quantum mechanics. Key aspects of Bohmian mechanics include:

Particle trajectories: Particles have definite positions and trajectories, guided by a "pilot wave" derived from the wave function.

Nonlocality: The pilot wave influences particle motion instantaneously, irrespective of distance.

Determinism: Although the initial conditions of particles might be unknown, the pilot wave fully determines their future behavior.

Critics of Bohmian mechanics argue that it lacks simplicity and elegance compared to other interpretations and may not be compatible with relativistic quantum theories.

2.4. Decoherence and Consistent Histories

The decoherence and consistent histories approach, developed by Wojciech Zurek, Murray Gell-Mann, and James Hartle in the 1980s and 1990s, offers a framework to understand the emergence of classical behavior from quantum systems. Key aspects of this approach include:

Decoherence: The interaction between a quantum system and its environment causes quantum coherence to be lost, effectively suppressing interference between different branches of the wave function.

Consistent histories: A set of coarse-grained histories that are mutually exclusive and exhaustive can be chosen, allowing for a classical description of the system's evolution.

No preferred interpretation: Decoherence provides a mechanism for wave function collapse-like behavior without invoking a specific interpretation of quantum mechanics.

Critics argue that decoherence alone does not solve the measurement problem, as it does not explain why a particular outcome is selected over others.

2.5. The Need for a New Perspective: Enter QBism

The ongoing debate around the interpretation of quantum mechanics has led to the exploration of alternative perspectives, one of which is Quantum Bayesianism, or QBism. QBism was developed in the early 2000s by physicists Christopher Fuchs, Carlton Caves, and Rüdiger Schack as an attempt to provide a coherent and consistent framework for understanding quantum mechanics. This approach addresses some of the criticisms and limitations of previous interpretations by emphasizing quantum probabilities' personal, subjective nature.

QBism is based on the Bayesian interpretation of probability, which treats probabilities as degrees of belief that are updated upon acquiring new information. In the context of quantum mechanics, QBism interprets the wave function and its probabilities as a representation of an observer's subjective knowledge about a quantum system, rather than an objective description of reality.

By adopting a subjective stance on probabilities, QBism sidesteps some of the conceptual issues associated with other interpretations, such as the nature of wave function collapse and the role of the observer. Moreover, it offers a fresh perspective on quantum paradoxes like entanglement and the EPR paradox, arguing that they stem from misconceptions about the role of probability in quantum mechanics.

QBism is not without its critics, who argue that its subjective nature undermines the objectivity of science and raises questions about the existence of an independent physical reality. Despite these concerns, QBism continues to stimulate new insights into the foundations of quantum mechanics and holds the promise of deepening our understanding of the quantum world.

3
A Bayesian Approach to Quantum Mechanics

3.1. The Basics of Bayesian Probability

Bayesian probability is a framework for reasoning about uncertain events by treating probabilities as degrees of belief, which are updated as new information becomes available. The foundation of Bayesian probability is Bayes' theorem, which relates the conditional probability of an event A given another event B to the probability of B given A:

$$P(A|B) = P(B|A) * P(A) / P(B)$$

In this context, $P(A|B)$ represents the updated or "posterior" probability of A after taking into account the new evidence B. $P(B|A)$ is the likelihood of observing B given A, while $P(A)$ and $P(B)$ are the prior probabilities of A and B, respectively.

3.2. The Subjective Nature of Probability

In contrast to the frequentist interpretation of probability, which treats probabilities as long-run frequencies of events, Bayesian probability views probabilities as subjective degrees of belief. This means that two individuals with different prior knowledge or beliefs may assign different probabilities to the same event. As they gather more data, their beliefs can be updated using

Bayes' theorem, eventually converging to a consensus as more information is accumulated.

3.3. Bayesian Inference and Its Applications

Bayesian inference is the process of updating beliefs in light of new data using Bayes' theorem. It has a wide range of applications, from scientific hypothesis testing and parameter estimation to machine learning and artificial intelligence. Some key advantages of Bayesian inference include its ability to incorporate prior knowledge, its capacity to deal with uncertainty, and its coherent framework for updating beliefs as new evidence is obtained.

3.4. How Bayesianism Informs Quantum Mechanics

Quantum Bayesianism, or QBism, applies the Bayesian approach to quantum mechanics by treating quantum states and probabilities as subjective degrees of belief. In this framework, the wave function is not an objective description of physical reality, but rather a representation of an observer's knowledge or beliefs about a quantum system.

By interpreting quantum probabilities as subjective, QBism offers a novel perspective on the measurement problem and wave function collapse. Instead of seeing wave function collapse as a physical process, QBism views it as an update of the observer's beliefs upon making a measurement. This reinterpretation has profound implications for our understanding of quantum mechanics and its various paradoxes, including entanglement, nonlocality, and the role of the observer.

While QBism has its critics, who argue that it threatens the objectivity of science, it nonetheless provides a unique and intriguing approach to the interpretation of quantum mechanics, shedding new light on some of its most perplexing features.

4
QBism: Principles and Foundations

4.1. Quantum States as Personal Degrees of Belief

A key tenet of QBism is that quantum states represent an observer's personal degrees of belief about the possible outcomes of a quantum measurement. This contrasts with other interpretations of quantum mechanics, which treat the wave function as an objective description of reality. By viewing quantum states as subjective, QBism places emphasis on the agent (the observer) and their knowledge about the system.

The subjective nature of quantum states in QBism allows for the possibility that two observers with different information might assign different quantum states to the same physical system. This is not seen as problematic because quantum states are treated as an agent's beliefs, rather than as an objective fact about the system.

4.2. The Role of the Observer

In QBism, the observer plays a central role in the understanding of quantum mechanics. The observer's beliefs and their subsequent updates upon measurement are crucial for making sense of quantum phenomena. This contrasts with other interpretations where the role of the observer is often seen as problematic, leading to questions about the nature of the measurement process and wave function collapse.

In QBism, the observer's role in measurement is not problematic because it is consistent with the Bayesian view of probability as subjective. When an observer performs a measurement, they update their beliefs about the quantum system in accordance with the Born rule, which is seen as a normative rule for updating one's beliefs, rather than an objective description of reality.

4.3. The Born Rule in QBism

The Born rule is a fundamental principle of quantum mechanics that relates the probabilities of measurement outcomes to the wave function. In QBism, the Born rule is interpreted as a rule for updating an observer's beliefs upon making a measurement.

This reinterpretation of the Born rule emphasizes the subjective nature of quantum probabilities and the central role of the observer. It also provides a fresh perspective on the measurement problem, as it no longer requires an explanation for wave function collapse as an objective physical process. Instead, the collapse is seen as an update of the observer's beliefs, in line with the Bayesian view of probability.

4.4. The QBist Framework for Quantum Mechanics

The QBist framework provides a consistent and coherent approach to quantum mechanics that emphasizes the subjective nature of quantum probabilities and the central role of the observer. By adopting a Bayesian approach, QBism addresses some of the longstanding issues and paradoxes associated with other interpretations of quantum mechanics, such as the measurement problem, entanglement, and nonlocality.

Key aspects of the QBist framework include:

1. Quantum states as subjective degrees of belief
2. The observer's central role in the measurement and belief updating
3. The reinterpretation of the Born rule as a normative rule for belief updating
4. While the QBist approach has its critics, it offers a fresh and intriguing perspective on quantum mechanics that may yield new insights and deepen our understanding of the quantum world.

5
Solving Quantum Paradoxes with QBism

5.1. The EPR Paradox and Nonlocality

The EPR paradox, formulated by Einstein, Podolsky, and Rosen in 1935, questions the completeness of quantum mechanics due to the seemingly nonlocal behavior of entangled particles. When two particles are entangled, the measurement of one particle's property appears to instantaneously influence the other particle's property, regardless of the distance between them.

QBism provides a different perspective on the EPR paradox by emphasizing the subjective nature of quantum probabilities. In QBism, entanglement represents a correlation between an observer's beliefs about the outcomes of measurements performed on the entangled particles. There is no need for a nonlocal influence or "spooky action at a distance," as the seemingly instantaneous correlation is a result of the observer updating their beliefs according to the Born rule.

5.2. Schrödinger's Cat and the Measurement Problem

Schrödinger's cat is a famous thought experiment illustrating the measurement problem in quantum mechanics, which questions how a definite outcome arises from the superposition of states in a wave function. In the experiment, a cat is placed in a box with a radioactive atom that, upon decay, triggers a mechanism that kills the cat. According to quantum mechanics, the cat exists in a superposition of alive and dead states until observed.

QBism resolves the Schrödinger's cat paradox by interpreting the wave function as a representation of an observer's beliefs rather than an objective description of reality. The cat is not in a superposition of states; instead, the observer's uncertainty about the cat's state is represented by the superposition. When the observer measures the cat's state, they update their beliefs in accordance with the Born rule, and the paradox dissolves.

5.3. The Role of Consciousness in Quantum Mechanics

The role of consciousness in quantum mechanics has been debated since the early days of the theory, with some interpretations suggesting that conscious observation is necessary for wave function collapse. QBism sidesteps this issue by treating quantum probabilities as subjective degrees of belief, making the question of consciousness less relevant.

In QBism, any observer, whether conscious or not, can update their beliefs upon making a measurement. There is no need for a special role of consciousness in the measurement process, as it is the observer's belief update that is central to the QBist interpretation.

5.4. QBism's Take on Many Worlds and Parallel Realities

The Many Worlds Interpretation (MWI) posits that every possible outcome of a quantum measurement occurs in separate, non-communicating branches of the universe. This has led to the idea of parallel realities, in which each branch represents a distinct reality where every possible outcome is realized.

QBism offers a different perspective on the notion of many worlds and parallel realities. In QBism, the wave function represents an observer's subjective beliefs about the possible outcomes of a measurement rather than an objective description of reality. As a result, there is no need to postulate multiple branches of the universe to accommodate different outcomes.

Instead, the observer updates their beliefs upon making a measurement, in line with the Born rule. This reinterpretation removes the need for parallel realities and avoids the complications that arise from the concept of multiple worlds.

6
Criticisms and Counterarguments

6.1. The Solipsistic Objection

A common criticism of QBism is that it appears to endorse solipsism, the idea that only one's own mind and experiences are certain to exist. By treating quantum states and probabilities as subjective degrees of belief, QBism might be interpreted as suggesting that physical reality is dependent on the observer's beliefs. However, QBism does not deny the existence of an external reality; instead, it emphasizes that our knowledge and predictions about that reality are necessarily subjective.

6.2. The Problem of Objectivity

Another criticism leveled at QBism is that its subjective nature undermines the objectivity of science. Scientific knowledge is often assumed to be objective and independent of any particular observer. By interpreting quantum mechanics in terms of subjective probabilities, QBism might be seen as a threat to the objectivity of scientific knowledge.

However, proponents of QBism argue that the interpretation maintains the objectivity of scientific practice by providing a coherent and consistent framework for understanding quantum mechanics, which ultimately leads to objective and verifiable predictions about the behavior of quantum systems.

6.3. Can QBism Be Tested?

Some critics argue that QBism, as an interpretational framework, is not testable and therefore cannot be considered a scientific theory. Because QBism deals with subjective probabilities, it might be difficult to design experiments that could definitively confirm or refute its claims.

In response, QBists contend that their interpretation provides a more coherent and consistent understanding of quantum mechanics, which can lead to new insights and predictions that could be tested in the future. Moreover, QBism's ability to resolve longstanding quantum paradoxes and conceptual issues suggests that it offers a valuable perspective on the foundations of quantum mechanics.

6.4. Responding to the Criticisms

Proponents of QBism acknowledge the criticisms but argue that the benefits of the interpretation outweigh its potential drawbacks. They maintain that QBism provides a coherent and consistent understanding of quantum mechanics, which can lead to new insights, predictions, and experimental possibilities. Furthermore, by resolving longstanding paradoxes and conceptual issues, QBism offers a fresh perspective on the quantum world that can deepen our understanding of its underlying principles.

Despite the criticisms, QBism continues to stimulate debate and research in the field of quantum mechanics, contributing to the ongoing quest for a deeper and more comprehensive understanding of the quantum world.

6.5. The Realism Debate

QBism has also been criticized for its apparent departure from scientific realism, the idea that science aims to provide an accurate and objective representation of an external reality. Critics argue that QBism's emphasis on subjective probabilities suggests that the quantum world is not truly independent of the observer, which conflicts with the realist view of science.

In response, QBists argue that their interpretation does not deny the existence of an external reality but merely focuses on the role of the observer in understanding and predicting that reality. Furthermore, they contend that QBism provides a more coherent and consistent account of quantum mechanics, which can help clarify the relationship between the observer and the external world.

6.6. The Problem of Communication

Another criticism of QBism concerns the issue of communication between observers. If quantum states represent subjective beliefs, critics argue that it may be difficult to reconcile the beliefs of different observers, potentially hindering scientific communication and collaboration.

QBists respond to this criticism by emphasizing that while quantum states are subjective, the predictions derived from those states can still be objectively compared and verified. This allows for effective communication and collaboration between observers, as they can compare their predictions and update their beliefs accordingly.

6.7. Responding to the Criticisms

Proponents of QBism acknowledge the criticisms but argue that the benefits of the interpretation outweigh its potential drawbacks. They maintain that QBism provides a coherent and consistent understanding of quantum mechanics, which can lead to new insights, predictions, and experimental possibilities. Furthermore, by resolving longstanding paradoxes and conceptual issues, QBism offers a fresh perspective on the quantum world that can deepen our understanding of its underlying principles.

Despite the criticisms, QBism continues to stimulate debate and research in the field of quantum mechanics, contributing to the ongoing quest for a deeper and more comprehensive understanding of the quantum world.

7
QBism in Practice: Applications and Future Developments

7.1. Quantum Computing and Information Theory

QBism's emphasis on the subjective nature of quantum states and probabilities has implications for quantum computing and information theory. By viewing quantum states as encoding an observer's knowledge about a system, QBism highlights the role of information in quantum mechanics. This perspective can inspire new algorithms, error-correction techniques, and other advances in quantum computing and communication by focusing on the observer's knowledge and how it can be manipulated and updated.

7.2. Quantum Cryptography and Security

Quantum cryptography, particularly quantum key distribution (QKD), relies on the principles of quantum mechanics to achieve secure communication. QBism's interpretation of quantum states as subjective degrees of belief can inform novel approaches to quantum cryptography, such as the design of new protocols that leverage the observer's subjective knowledge to ensure security. Understanding the observer's role in the process of measuring and updating beliefs about quantum systems can provide new insights into how information can be securely transmitted and protected from eavesdropping.

7.3. Quantum Foundations and Conceptual Advances

As an alternative interpretation of quantum mechanics, QBism can help deepen our understanding of the theory's foundations and lead to conceptual advances. By providing a fresh perspective on longstanding paradoxes and conceptual issues, QBism can stimulate new lines of research and exploration into the nature of quantum systems. This may result in the discovery of new principles, theorems, or relationships that can further advance our understanding of quantum mechanics.

7.4. The Future of QBism

QBism has already generated significant interest and debate within the field of quantum mechanics. Its potential to reshape our understanding of the quantum world and inspire new applications in quantum computing, cryptography, and other areas make it a promising area of research for the future.

As the field of quantum mechanics continues to evolve, QBism may provide new insights and solutions to outstanding problems or even give rise to new questions and challenges. The ongoing development of QBism, its refinement, and its integration with other areas of quantum research will undoubtedly contribute to the broader effort to deepen our understanding of the quantum world and harness its unique properties for practical applications.

8
Competing Emergent Theories in Quantum Mechanics

As quantum mechanics continues to evolve, several emergent theories have been proposed to address the interpretational challenges and paradoxes associated with the field. These theories, like QBism, seek to provide a fresh perspective on quantum mechanics and offer new solutions to longstanding issues. This chapter will explore some of these competing emergent theories and their implications for our understanding of the quantum world.

8.1. Relational Quantum Mechanics

Relational Quantum Mechanics (RQM) is an interpretation that posits that quantum states are relative to the observer, emphasizing the relationships between different physical systems. In RQM, the properties of a system are only defined concerning another system, and there is no absolute or objective description of the quantum world. While RQM shares some similarities with QBism in terms of subjectivity, it focuses more on the relational aspects of quantum states rather than their probabilistic nature.

8.2. Quantum Darwinism

Quantum Darwinism is a theoretical framework that attempts to explain the emergence of classical, objective reality from the quantum world. According to this approach, information about a quantum system is "selected" and proliferated throughout the environment via a Darwinian process, leading to the creation of multiple redundant copies of the information. This process ultimately results in the emergence of a classical reality, as the information becomes effectively objective and widely accessible to observers.

8.3. Quantum Bayesian Networks

Quantum Bayesian Network (QBN) is a graphical model that combines quantum mechanics with Bayesian probability theory. This approach provides a unified framework for modeling and reasoning about quantum systems and their interactions with classical systems. QBNs have the potential to bridge the gap between quantum and classical worlds by offering a more intuitive and visual representation of quantum processes, allowing researchers to explore complex quantum phenomena more effectively.

8.4. Superdeterminism

Superdeterminism is an alternative explanation for the apparent nonlocality observed in entangled quantum systems. This theory posits that all events in the universe are predetermined and that the outcomes of quantum experiments are influenced by hidden variables that were determined at the universe's inception. Superdeterminism, though controversial, offer a potential resolution to the EPR paradox and Bell's theorem by proposing that the correlations observed in entangled systems are due to a pre-established, deterministic cosmic order.

These emergent theories and QBism contribute to the ongoing debate and research into the interpretation of quantum mechanics. Each theory offers unique perspectives on the quantum world, addressing different aspects of the interpretational challenges and paradoxes that have long perplexed physicists. By exploring these alternative approaches, researchers continue to expand our understanding of the quantum realm and its connections to the classical world we experience.

Conclusion
Quantum Mechanics
in a Personal Light

QBism offers a novel and thought-provoking perspective on quantum mechanics by casting it in a personal light. By treating quantum states and probabilities as subjective degrees of belief, QBism emphasizes the role of the observer in understanding and predicting the behavior of quantum systems. This approach stands in contrast to more traditional interpretations, which view quantum states as objective descriptions of reality.

The QBist framework provides a consistent and coherent interpretation of quantum mechanics that can help resolve longstanding paradoxes and conceptual issues, such as the measurement problem, entanglement, and nonlocality. QBism dissolves many of the mysteries that have plagued quantum mechanics since its inception by interpreting the wave function as a representation of an observer's beliefs.

QBism has the potential to impact various applications, such as quantum computing, information theory, and cryptography, by providing a fresh perspective on the role of information and the observer's knowledge. By focusing on the subjective nature of quantum mechanics, QBism can inspire new approaches and techniques that exploit the observer's beliefs to develop more efficient and secure technologies.

Despite its criticisms and challenges, QBism has generated significant interest and debate within the quantum mechanics community. As a result, it has the potential to stimulate new lines of research, foster conceptual advances, and contribute to a deeper and more comprehensive understanding of the quantum world.

QBism invites us to view quantum mechanics in a personal light, emphasizing the subjective nature of quantum states and probabilities and the central role of the observer. Doing so provides a unique and intriguing approach to interpreting quantum mechanics that may ultimately enrich our understanding of the quantum world and its many mysteries.

Reference

QBism: The Future of Quantum Physics *by* Hans Christian Von Baeyer